Negative Numbers – A Collection of Exercises

Stephan Sigler

edition waldorf

Heartfelt thanks to Benedikt Kaiser and Peter Schwab for their help with the layout, and to Peter also for the numerous corrections and suggestions, as well as to the members of the mathematics curriculum group of the Pedagogical Research Institute of the Association of Waldorf Schools for their support and critical collaboration.

English translation by Charles Gunn.
Editor of the english version: Robert Neumann.

Impressum

Bildungswerk Beruf und Umwelt
Brabanter Str. 30 | 34131 Kassel
Telefon 0561-37206 | Fax 0561-3162189
www.lehrerseminar-forschung.de
info@lehrerseminar-forschung.de
© Bildungswerk Beruf und Umwelt
1. Edition 2016

ISBN 978-3-939374-28-2

This book and the associated teacher's book with pedagogical information can be ordered online:
www.lehrerseminar-forschung.de | www.waldorfbuch.de

Contents

1 Exercises for up and down

1.1 Account transactions

1. Credit – debit

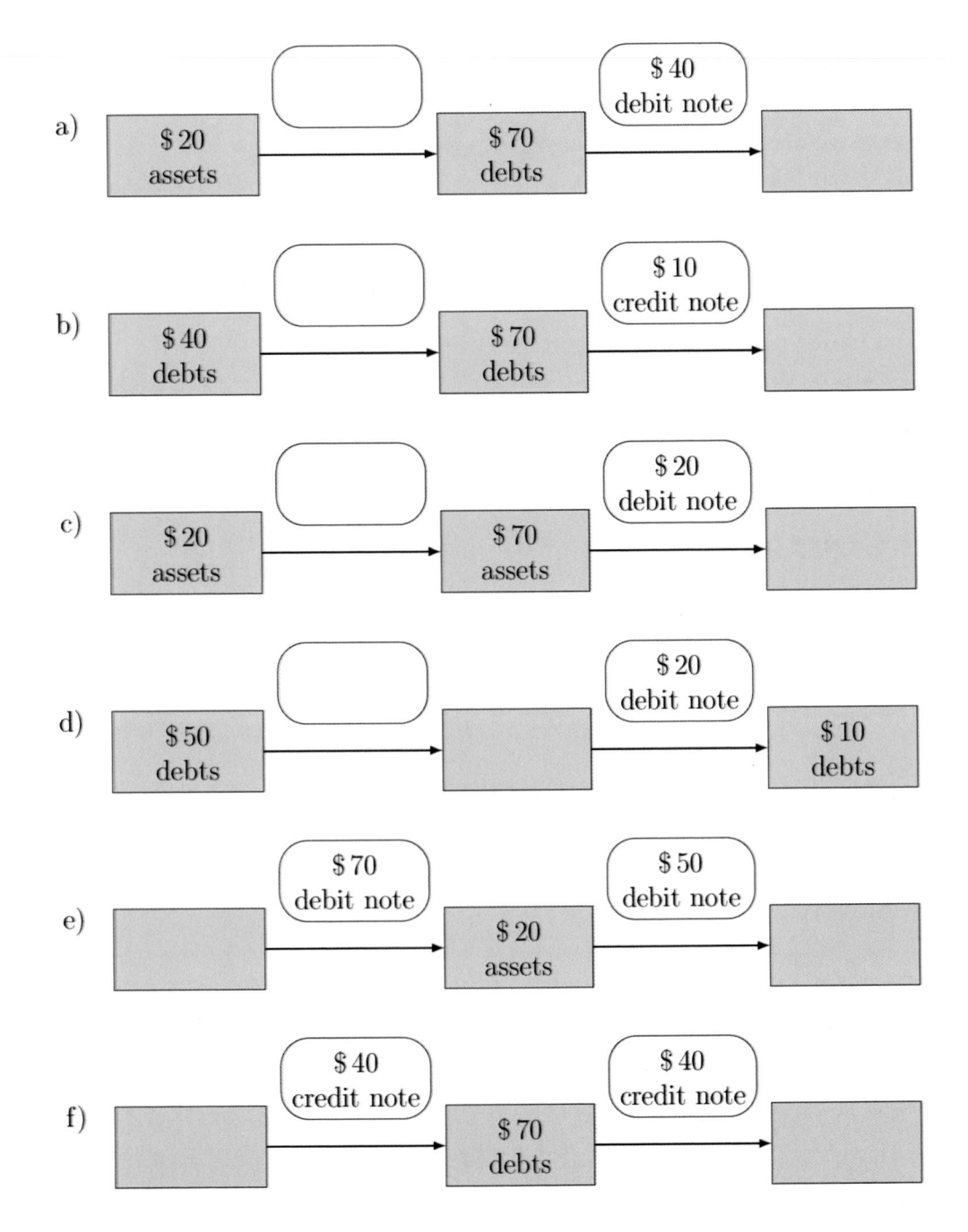

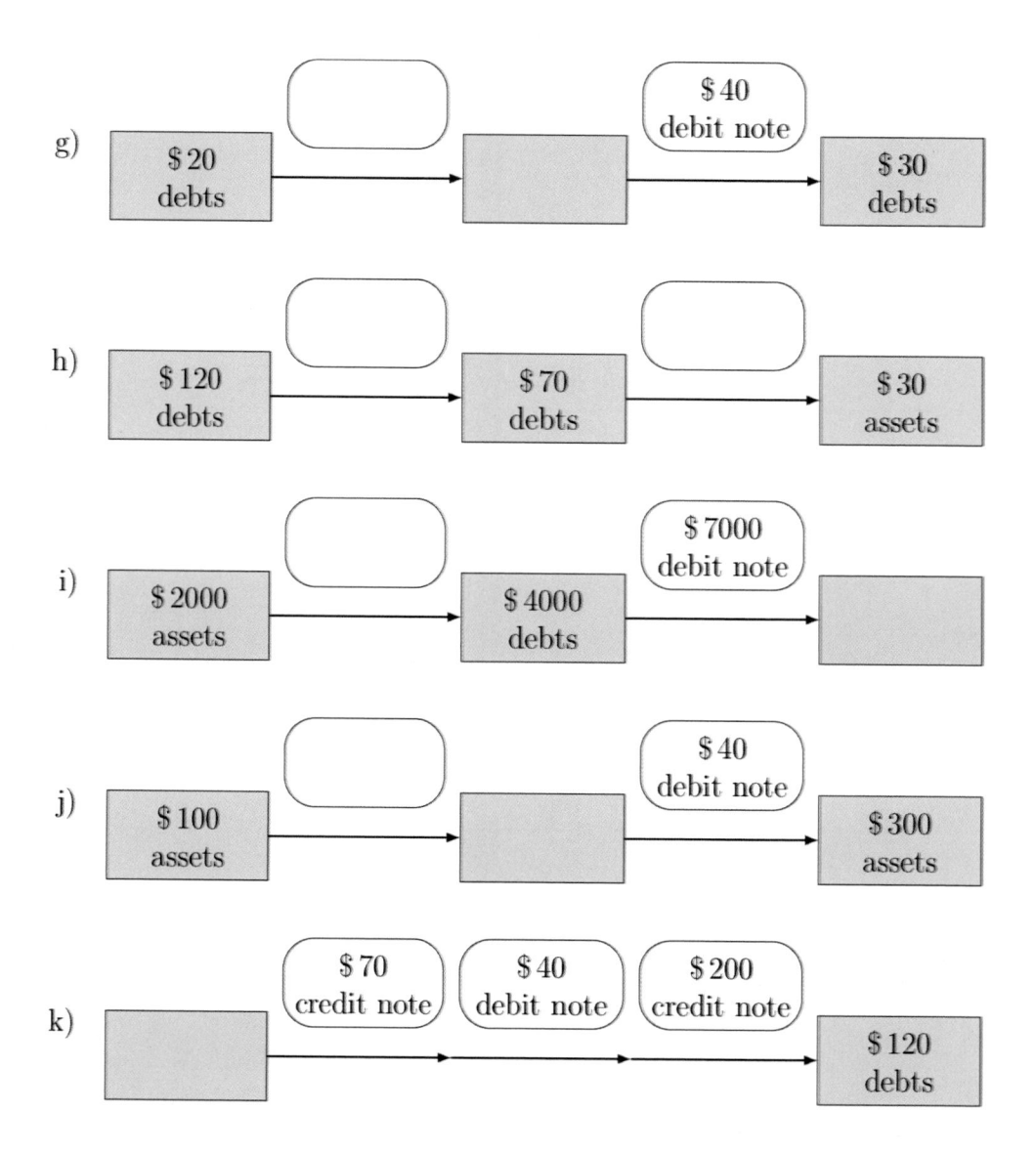

2. You can write down account transactions as follows: CN = credit note, DN = debit note. For account balances, indicate debts with a − and assets with a +.

Credit/debit note in $	Account balance in $
	1200+
800 CN	2000+
2200 DN	200−

Please fill in the blanks in the following accounts:

Credit/debit note in $	Account balance in $
	100+
	140+
	100−
	50−

Credit/debit note in $	Account balance in $
	1000+
120 CN	
20 DN	
200 CN	

Credit/debit note in $	Account balance in $
	125+
75 DN	
50 DN	
	241−

Credit/debit note in $	Account balance in $
	125+
	120−
150 CN	
	225+

3. Be careful, the numbers here are more difficult. If you can't work out the answers in your head, then calculate them in your exercise book.

Credit/debit note in $	Account balance in $
	123−
243 DN	
57 CN	
111 CN	

Credit/debit note in $	Account balance in $
	27,50+
12.20 CN	
47.30 DN	
12.90 DN	

Credit/debit note in $	Account balance in $
	125.50+
147.80 DN	
50.40 CN	
	41.40+

Credit/debit note in $	Account balance in $
	24.90+
50.40 CN	147.80−
	41.40+

4. In the bank statements of financial institutions, assetes and credit notes are marked with a +, debt and debit are on the other hand marked with a − after the amount. Fill in the blanks.

Text	Amount in $
old balance	200+
debit note	300−
credit note	700+
debit note	1500−
new balance	

Text	Amount in $
old balance	1700+
debit note	1300−
debit note	500−
credit note	200+
new balance	

Text	Amount in $
old balance	250−
credit note	350+
debit note	850−
credit note	1800+
new balance	

Text	Amount in $
old balance	222−
debit note	148−
debit note	456−
credit note	259+
new balance	

Can one combine the three account transactions into one credit or debit note? What amount would this be?

5. Mr. Meier has five accounts. The account balances are:

Account	Account balance
A	$ 1250.41 credit
B	$ 245.31 credit
C	$ 1452.01 credit
D	$ 2.65 credit
E	$ 89.99 debit

Does Mr. Meier have more assets or debts? What is the total balance?

6. Another exercise involving assets and debts. Abbreviate assets with A and debts with D.

   a) $ 72.45 debts and $ 123.23 credit result in _____ .

   b) $ 245.12 credit and $ 1146.78 debts result in _____ .

   c) $ 3454 debts and $ 123,678.78 debts result in _____ .

   d) $ 122 debts and $ 234 credit and $ 134 debts result in _____ .

   e) $ 365 credit and $ 112 debts and $ 13 debts and $ 1000 debts result in

   _____ .

   f) $ 23.56 debts and $ 123.57 credit and $ 75.23 credit and $ 175.24 debts result in _____ .

1.2 Mixed exercises

In some countries the ground floor of a building is denoted by the number "0". So if you want to get out of the building, you presses the "0" button in the elevator.

1. There is a elevator built in a multi-storey building which has 20 floors for apartments and 10 basement car parking floors.

   a) Mr. A gets in the elevator on the 16th floor and goes down to the 4th basement floor. How many floors has he passed?

   b) Mr. B goes up 12 floors and gets out on the 9th floor. On which floor did he get on?

   c) Mr. C goes down 17 floors and got in on the 3th floor. On which floor does he get out?

   d) Mr. D enjoys going in the elevator. He goes up 7 floors and then 11 more, after that he goes down 15 floors, then again 5 floors upwards. On the 3rd floor he finally gets out. On which floor did he get on the elevator?

e) Create exercises similar to the previous ones and give them to your neighbour to solve.

f) Write down how you devised the exercises as well as how to solve them.

2. Peter bought a new bike. His parents gave him $250 as a present towards the bike. Since the bike he wanted cost more than that, Peter had to borrow money from his parents. For the past five months he has been paying back $4 monthly from his pocket money back. His grandmother also gave him $20 towards the bike. Peter now has $49 of debt left to pay. How much did the bike cost? How many more months does he have to pay back?

3. On one day there are the following temperatures in different cities:

Moscow	−15°C	London	4°C
Irkutsk	−26°C	Frankfurt	2°C
Rome	12°C	Berlin	−3°C
Athens	13°C	Helsinki	−6°C

a) Which 2 cities have the biggest difference in temperature? How high is this difference?

b) Write down (at least) 5 sentences that start like this: "Compared to London, it is 19°C colder in Moscow".

c) Now you can look up these places in the atlas.

4. On Tuesday in the three cities Seville, Oslo and Kassel it is 2.7°C colder than on Monday. On Wednesday the temperature goes up 6.4°C, on Thursday it goes up again 1.3°C. Please complete the following table showing the measured temperature at 12:00 p.m.

	Monday	Tuesday	Wednesday	Thursday
Seville	12.3°C			
Oslo	−8.5°C			
Kassel	−1.8°C			

5. Figure out what time it is with the help of the time zones given in the atlas:

	Problem	Solution
a)	What is the time in Sydney when it´s 12 p.m. in Greenwich?	
b)	What is the time in Johannesburg when it´s 2 p.m. in Greenwich?	
c)	What is the time in Sydney when it´s 6 a.m. in Moscow?	
d)	What is the time in Shanghai when it´s 6 p.m. in Tokyo?	
e)	What is the time in Djakarta when it´s midnight in Auckland?	
f)	What is the time in Athens when it´s 9 a.m. in Berlin?	
g)	What is the time in Sydney when it´s midnight in New York?	
h)	What is the time in Mexico City when our school starts?	
i)	What is the time in Moscow when it´s 11 a.m. in San Francisco?	
j)	What is the time in Lima when it´s 11 p.m. in Delhi?	
k)	What is the time in Honolulu when it's 11 a.m. in Rome?	

6. Using am for times between midnight and noon and pm for times noon and midnight, what date and time was it in...

   a) Honolulu – when it was 0:05 a.m. in 12.05.2003 in Rome?

   b) Ottawa – when it was 4:00 a.m. in 12.24.2003 in Berlin?

   c) Sydney – when it was 10:00 p.m. 12.31.2003 in Teheran?

   d) Berlin – when it was 0:01 a.m. the 01.01.2000 in Sydney?

1.3 Lakes on earth

In order to orientate ourselves on the earth's surface, we need in many situations to know the difference in altitude between two particular places. For this purpose, people have agreed to specify such altitudes with respect to the level of the world's oceans, the so-called "sea level". Then one normally speaks – for example – of an altitude of 357 meters above sea level (ASL). We denote levels below "sea level" as BSL. If you know the altitudes of two places in this form, then you can also calculate the difference of their altitudes. In the following exercise you are first to enter the information given for the various lakes into the table. Then, fill in any remaining blank spaces by calculating.

1. The water level of the Lake Baikal lies 455 m ASL; the sea bottom however is 1165m below sea level (BSL). This is the deepest lake on earth.

2. The Dead Sea (Israel/Jordan) lies in a deep channel set into the earth's surface, so that the sea level of the Dead Sea is 400m BSL. The water depth is 396 m.

3. The Caspian Sea has a surface of 386,400 km covering an area even bigger than Germany. Its water level is 371 m higher than the water level of the Dead Sea. The deepest point on the sea bottom is 1023 m BSL.

4. The Great Lakes of North America form the largest connected fresh-water area on earth. Lake Superior is the biggest of the five. It has a depth of 397 m. The water surface lies 183m ASL.

5. The longest lake in Africa is Lake Tanganyika which is however one-tenth the size of the Caspian Sea. It has a depth of 1470 m and its sea bottom lies 688 m BSL.

Lake	Water level	Lake floor	Water depth
Lake Baikal	455 m.a.s.l.	1165 m.b.s.l.	
Dead Sea			
Caspian Sea			
Lake Superior			
Lake Tanganyika			

See whether you can find these lakes in the atlas.

1.4 Temperature records

At different places and times on the earth, temperature records have been measured. (The temperatures are given in °C).

1. The highest temperature recorded in Europe is 50.0°C in Seville.

2. This temperature is 8°C below the highest temperature measured in Libya.

3. In Death Valley the highest record was 56.7°C. This temperature lies 31.7°C above the yearly average for Death Valley.

4. The lowest temperature measured was in the Russian Antarctic at Wostok, 138°C below the highest temperature measured in Seville.

5. The monthly average in winter for Wostok is 60°C below zero. The average summer temperature is 35°C higher.

6. It is not as cold in Greenland, where the lowest temperature was 22°C above the lowest temperature in Wostok.

7. The coldest place on earth is Oimjakon in North Siberia. There 77.8°C below zero has been measured. The highest temperature in summer was 108°C higher than this.

8. The normal seasonal weather in Oimjakon is as follows: The coldest months are December and January, temperatures are around -48°C. By July they rise 64°C above the coldest measurements.

9. The highest temperature measured was in Dallol in Ethiopia, 9.4°C above the highest temperature in Death Valley.

Please fill in the blanks that are not in gray. Look up more information about the location of these places in the atlas as well as any additional interesting facts.

Place	Lowest Value	Average Value		Highest Value
Seville				
Libya				
Death Valley				
Wostok/Antarctica		winter:	summer:	
Greenland				
Oimjakon		Dec/Jan:	July:	
Dallol/Ethiopia				

1.5 Calculating the game of darts

When playing darts on the dart board shown in the figure, you receive between 10 points – for hitting the small center circle – and 1 point for hitting the circle third from the outside. If you throw the dart into the two outer rings, or miss hitting any ring, then you receive 1, 2 or 5 negative points, *resp.* In every round, the player throws three darts. The sum of the three throws is recorded.

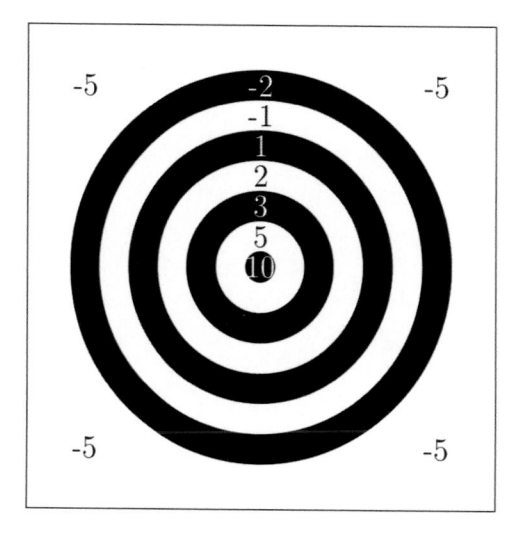

1. Calculate the five best possible results in one round as well as the five worst possible results. Please fill in each throw and the results in the table below.

result		1st round		2nd round		3rd round
$(+30)$	$=$	$(+10)$	and	$(+10)$	and	$(+10)$
	$=$		and		and	
	$=$		and		and	
	$=$		and		and	
	$=$		and		and	
(-15)	$=$	(-5)	and	(-5)	and	(-5)
	$=$		and		and	
	$=$		and		and	
	$=$		and		and	
	$=$		and		and	

2. Think about wheather all results between -10 and 10 are possible within a round of three turns. Write down the possible results below. Find as many possibilities as you can for each result!

2 Adding positive and negative numbers

2.1 Introductory exercises

1. Fill in the missing numbers in the equations:

a) ☐ plus (-3) = $(+5)$

b) ☐ plus $(+5)$ = $(+2)$

c) (-15) plus ☐ = (-4)

d) $(+23)$ plus ☐ = $(+7)$

e) (-5) plus ☐ = (-12)

f) ☐ plus $(+14)$ = $(+32)$

g) (-3.5) plus ☐ = 0

h) ☐ plus $(+4.5)$ = (-10.5)

i) ☐ plus (-3.1) = (-3.1)

j) $(+2.4)$ plus ☐ = $(+2.6)$

k) (-3.75) plus ☐ = (-1.25)

l) ☐ plus (-2.9) = $(+12)$

m) ☐ plus $(+2.3)$ = (-12.3)

n) $(+2.25)$ plus ☐ = $(+4.5)$

o) $(+2.12)$ plus ☐ = 0

2. Try to do these in your head. If you can´t manage it, then use an sheet of paper.

a) (+12) plus (−23) = _____

b) (+120) plus (−43) = _____

c) (−4) plus (+65) = _____

d) (−120) plus (−43) = _____

e) (−23) plus (−23) = _____

f) (+2.2) plus (−5.4) = _____

g) $\left(+\frac{3}{4}\right)$ plus $\left(-\frac{7}{4}\right)$ = _____

h) $\left(-\frac{7}{3}\right)$ plus $\left(+3\frac{6}{7}\right)$ = _____

i) (−35) plus (−3.1) = _____

j) (+2.4) plus (−123) = _____

k) (−76) plus (+45.7) = _____

3. Fill in the missing signs. Sometimes there are two possibilities!

a) (29) plus (18) = (−11)

b) (13) plus (−25) = (38)

c) (+14) plus (6) = (8)

d) (17) plus (13) = (−30)

e) (27) plus (15) = (+42)

f) (+19) plus (49) = (30)

g) (20) plus (−13) = (33)

h) (90) plus (30) = (120)

i) (15) plus (−6) = (21)

    j)   (+12)   plus   (8)   =   (20)

    k)   (−49)   plus   (53)   =   (102)

    l)   (14)   plus   (5)   =   (−19)

    m)   (42)   plus   (+28)   =   (14)

    n)   (129)   plus   (72)   =   (57)

4. Try to add as efficiently as possible:

    a) (+12) plus (−7) plus (+18) plus (−3) =

    b) (−1.5) plus (+2.3) plus (+2.4) plus (−2.3) plus (+7.5) plus (+2.5) =

5. Look at the following calculation:

$$(-12) \text{ plus } (+8) = (-4)$$

It consists of two terms (−12) and (+8) which equal (−4). What happens to the second term when the first term increases by 1; 2; 3... and the answer remains (−4)?

6. This is all that remains of an exercise:

The exercise involved numbers of magnitute 2 and 5 and 7 (either positive or negative). Unfortunately they were erased. How might this calculation have looked like? How many solutions can you find? One example is $(+7)$ plus $(-2) = (+5)$

7. Which of the following additions produces the largest sum? Think about it first without calculating, then calculate the sums and consider how you can find the largest sum without working everything out.

$$(-173) \text{ plus } (+458) \quad =$$
$$(-174) \text{ plus } (+459) \quad =$$
$$(-172) \text{ plus } (+459) \quad =$$
$$(-174) \text{ plus } (+457) \quad =$$

2.2 Addition walls

1. Fill in the empty bricks. The content of each brick is the sum of the numbers in the two bricks it is resting on.

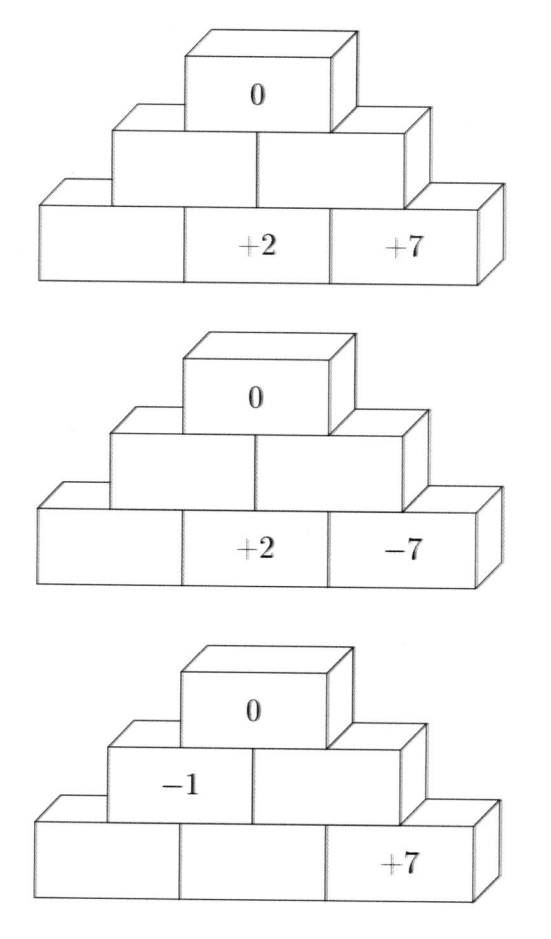

In the following addition wall, you have to first consider and try out the different possibilities! Finding the answer is not so easy!

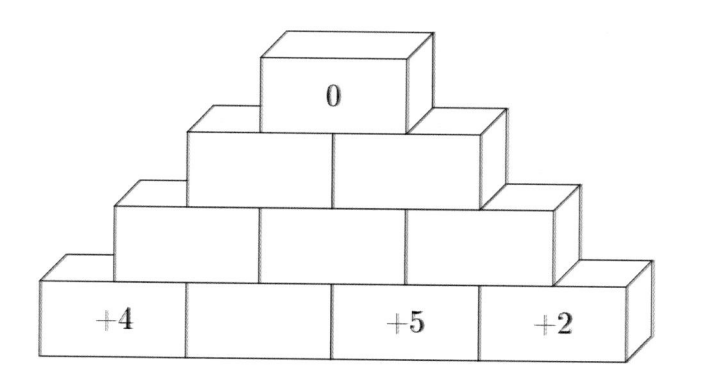

2.3 Something magic

1. Magic squares:
   Numbers can be arranged in magic squares with 4 rows and 4 columns. Every number should only appear once. If you calculate the sum of every row and column in these "magic squares" you will see that the answer will always be the same. Even the diagonal numbers have the same "magic" answer.

   a) Now check whether the following squares give you a magic number.

−7	6	7	−4
4	−1	−2	1
0	3	2	−3
5	−6	−5	8

−7	4	3	−10
−2	−5	−4	1
−6	−1	0	−3
5	−8	−9	2

   b) In this square the numbers between 1 and 16 should appear. The "magic number" in this case is 34. Fill in the missing numbers in order to get to 34. Remember, you can´t use any number twice:

1	14	15	4
	7		
	11	10	5

   c) Think about why the "magic number" in the last magic square had to be 34, when the numbers from 1 to 16 each occur once. Also check the magic number in the exercise in 1a).
   Here is a clue: add together all the numbers appearing in the magic square.

2. **Magic crosses**

Fill in the blanks so that the horizontal and vertical sum is the same. There are several possible answers. You should give at least five answers. How do the vertical and horizontal numbers you have put in the blanks relate to each other?

First cross (left):

		+3				
		−5				
+3	−4		+2	−1	+11	−5
		+1				
		−2				
		−10				

Second cross (right):

		−15				
		+7				
+124	−57		+2	−67	+12	−5
		+87				
		−21				
		−10				

2.4 Further exercises

1. How far apart are these numbers from each other?

  a) +6 and +18   ☐   d) −8 and +4   ☐

  b) −6 and +18   ☐   e) −2 and +38,5   ☐

  c) −6 and −6   ☐   f) −14 and −37   ☐

2. Fill in the number which is exactly mid-way between the two given numbers:

  a) +6 and +18   ☐   d) −8 and +4   ☐

  b) −6 and +18   ☐   e) −2 and +18   ☐

  c) −6 and +6   ☐   f) −17 and +22   ☐

3. When looking at the number -2, there are two numbers at a distance of 5 from it: $+3$ and -7.
   Which numbers have

   a) the distance of 4 from the number $+7$ ⬚ and ⬚

   b) the distance of 4 from the number $+2$ ⬚ and ⬚

   c) the distance of 7 from the number 0? ⬚ and ⬚

   d) the distance of 10 from the number $+3$? ⬚ and ⬚

   e) the distance of 187 from the number -234? ⬚ and ⬚

   f) the distance of 507 from the number -384? ⬚ and ⬚

4. Continue the number sequences with three further terms:

   a) $+7, +5, +3, \ldots$ ⬚⬚⬚

   b) $-10, +9, -11, +10, \ldots$ ⬚⬚⬚

   c) $-19, -14, -9, \ldots$ ⬚⬚⬚

   d) $-11, -5, 0, +4, \ldots$ ⬚⬚⬚

   e) $+7, +5, +4, +2, \ldots$ ⬚⬚⬚

   f) $+18, +17, +15, +11, +3, \ldots$ ⬚⬚⬚

   g) $+2, +3, +1, +2, -1, 0, -4, \ldots$ ⬚⬚⬚

   h) $+1, +3, -2, -6, +4, +12, \ldots$ ⬚⬚⬚

5. Consider and enter the result in the space provided:

   a) Is -3.4 closer to -3 or to -4? ⬚

   b) Is -8.6 closer to -9 or to -8? ⬚

   c) Is -4.5 closer to -2.7 or to -6.4? ⬚

   d) Is $-\frac{3}{4}$ closer to -1.5 or to 0? ⬚

3 Subtracting positive and negative numbers

3.1 Introductory exercises

1. Calculate the answer and fill in the next three rows:

$(+13)$ minus $(+7) =$ (-13) minus $(+5) =$

$(+13)$ minus $(+9) =$ (-13) minus $(+1) =$

$(+13)$ minus $(+11) =$ (-13) minus $(-3) =$

_____

_____

_____

(-13) minus $(+2) =$ $(+13)$ minus $(-1) =$

(-13) minus $(+5) =$ $(+13)$ minus $(-5) =$

(-13) minus $(+8) =$ $(+13)$ minus $(-9) =$

_____

_____

_____

2. Calculate:

a) $(+13)$ minus $(+3) =$ c) (-8) minus $(+12) =$

b) $(+13)$ minus $(+17) =$ d) (-1) minus $(+20) =$

3. Fill in the blanks.

a) $(+2)$ plus (-5) minus $(\quad) =(+2)$

b) $(+2)$ plus (-15) minus $(\quad) = (+2)$

c) $(\quad)$ plus (-6) minus $(-6) = (+2)$

d) $(+2)$ plus $(\quad)$ minus $(+10) = (+2)$

e) $(\quad)$ plus (-5) minus $(-5) = (+12)$

4. Fill in the blanks. The two exercises in each row are related to each other. How?

$(+6)$ plus $($ $) = (+4)$ $($ $)$ minus $(+2) = (+4)$

$($ $)$ plus $(+19) = (+23)$ $(+23) = (+4)$ minus $($ $)$

$($ $)$ plus $(+8) = (+25)$ $(+17)$ minus $($ $) = (+25)$

$(-12) = ($ $)$ minus (-3) $($ $) = (-15)$ plus $(+3)$

(-7) plus $($ $) = (-9)$ (-7) minus $(+2) = ($ $)$

$($ $)$ plus $(+8) = (+3)$ (-5) minus $(-8) = ($ $)$

5. Calculate:

a) $(+7)$ minus $(+7) = $ _____   g) (-5) minus $(+3) = $ _____

b) $(+3)$ minus $(+3) = $ _____   h) (-5) minus $(+7) = $ _____

c) (-7) minus $(-7) = $ _____   i) $(+5)$ minus $(-3) = $ _____

d) (-3) minus $(-3) = $ _____   j) $(+5)$ minus $(-7) = $ _____

e) $(+5)$ minus $(+3) = $ _____   k) (-5) minus $(-3) = $ _____

f) $(+5)$ minus $(+7) = $ _____   l) (-5) minus $(-7) = $ _____

6. Fill in the blanks.

a) $(+2) = ($ $)$ minus (-4) f) $(+5) = ($ $)$ minus (-8)

b) $(+2) = ($ $)$ minus (-6) g) $(-3) = ($ $)$ minus (-4)

c) $(+2) = ($ $)$ minus (-8) h) $(-3) = ($ $)$ minus (-7)

d) $(+5) = ($ $)$ minus (-4) i) $(-3) = ($ $)$ minus (-10)

e) $(+5) = ($ $)$ minus (-6)

3.2 Mixed exercises

1. The following numbers are given: -1271, $+1498$, -1765, $+3374$. You can take two of the four numbers and subtract them from each other for example: (-1271) minus $(+1498)$.

   a) Which two numbers produce the largest possible number when you subtract one from the other? Try it out!

   b) Which two numbers produce the smallest possible number when you subtract one from the other? Try it out!

2. The following four number towers in a row are constructed according to a specific calculation principle. In order to get the number in the block to the right, you have to carry out a specific calculation involving the two numbers and a third number not shown. Describe what this calculation is.

10	
4	16

-15	
3	-10

-12	
-10	-20

9	
3	14

3. The following number towers use the same concept of calculation as the previous exercise. What is the missing number?

12	
7	

-12	
-1	

6	
-6	

5	
	0

5	-11

-10	
	14

4. Now it is your turn to create your own number tower. Please use negative numbers as well and ask your neighbour figure out the principle you have used.

5. Read the following exercise very carefully and then calculate the answer. Write down the full calculation.

   a) Which number must I subtract from (+48) to get (+60)?

   b) Which number must I subtract from (+15) to get (+26)?

   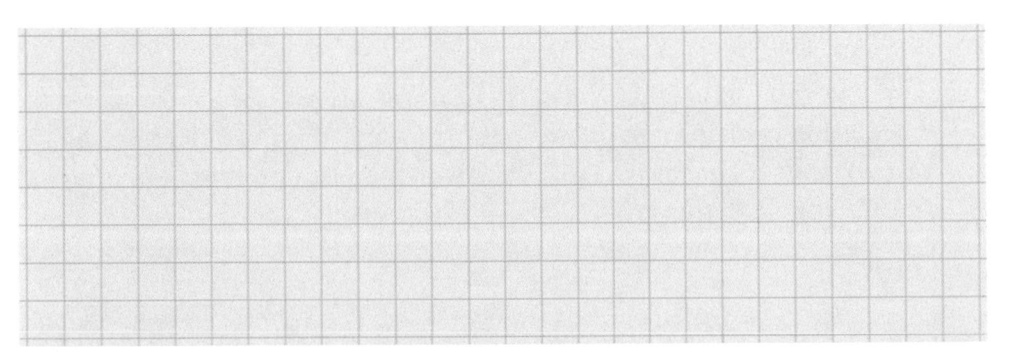

   c) Which number must I subtract from (-57) to get (+3)?

6. When you subtract two numbers from each other, the result is called the difference. For example when you subtract 4 from 7 you will get 3 as the difference. Can one say that the difference of these two numbers is also the distance between them? Now calculate more examples with positive numbers, then with positive and negative numbers and then only with negative numbers. Formulate a conclusion.

7. Addition walls: Add and subtract alternating, based on the operation shown to the left of each row of bricks.

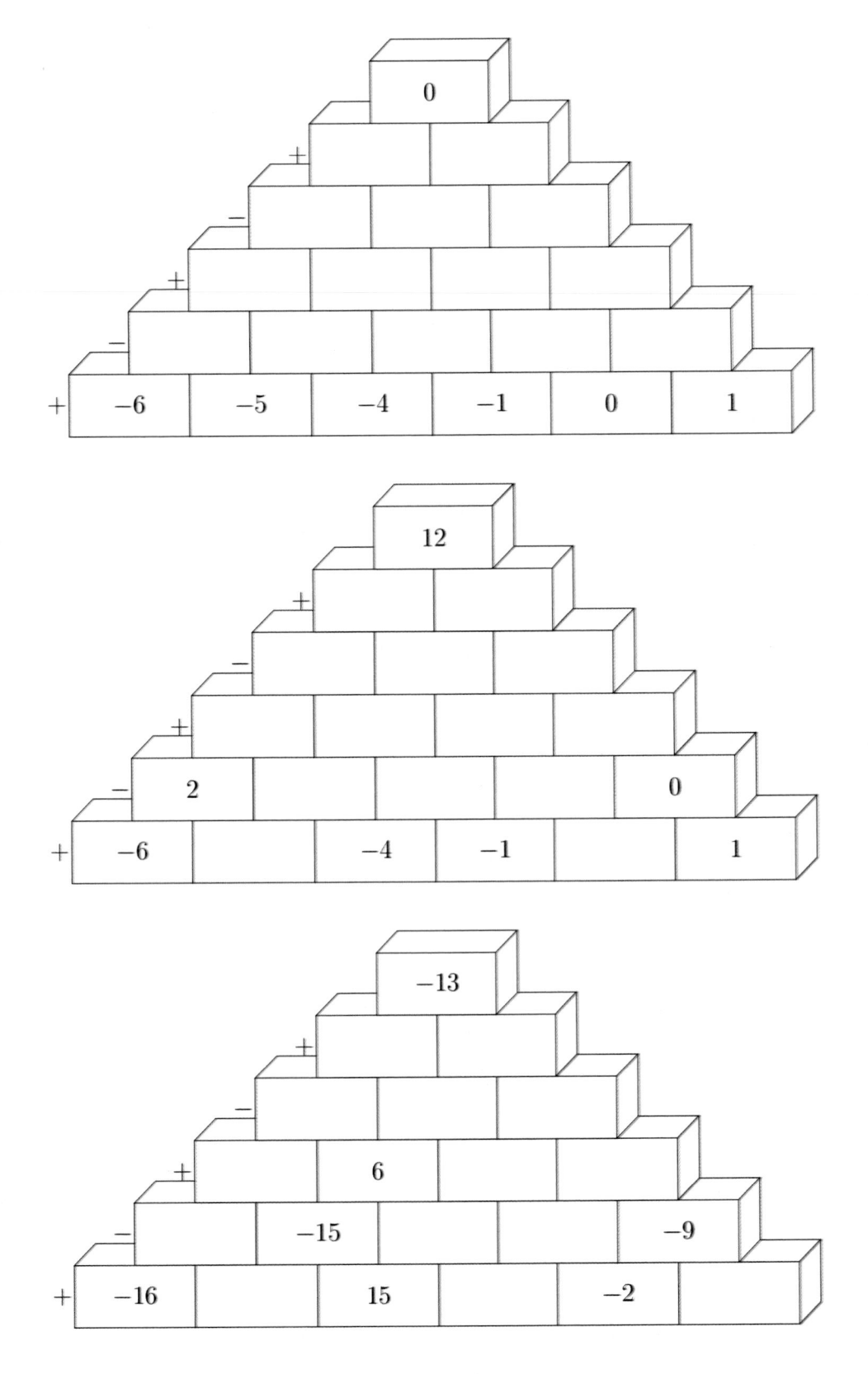

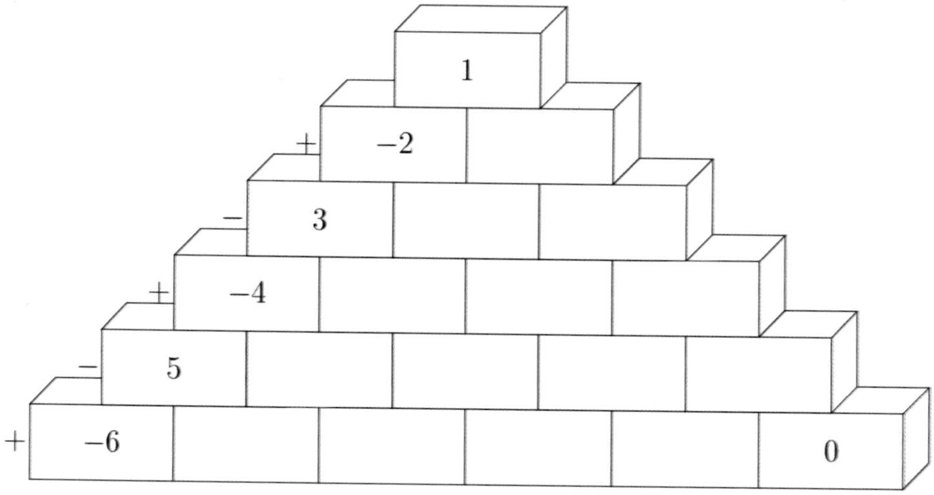

1
+ −2
− 3
+ −4
− 5
+ −6
0

11
+ −8
− −28
+ −14
− −15
+ −17
49

In the following exercise, using only addition, devise a solution so that the numbers in the last row are all the same.

−192
+
+
+
+
+

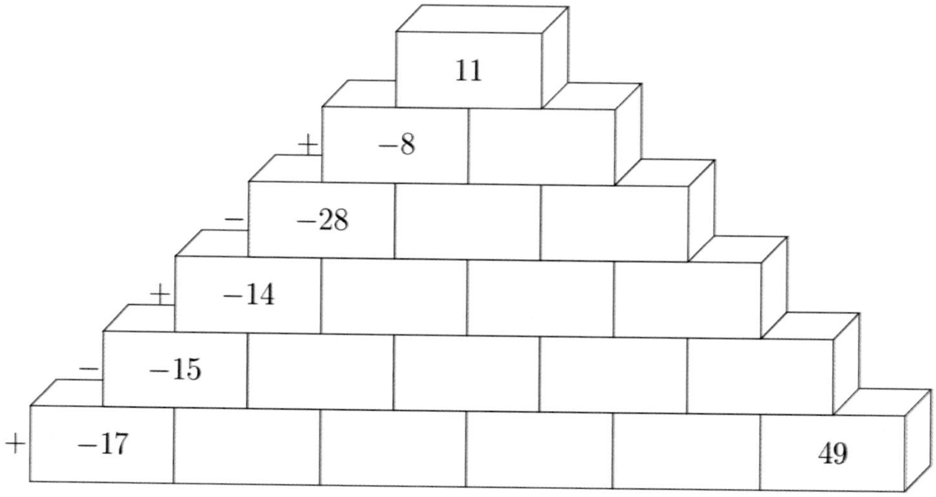

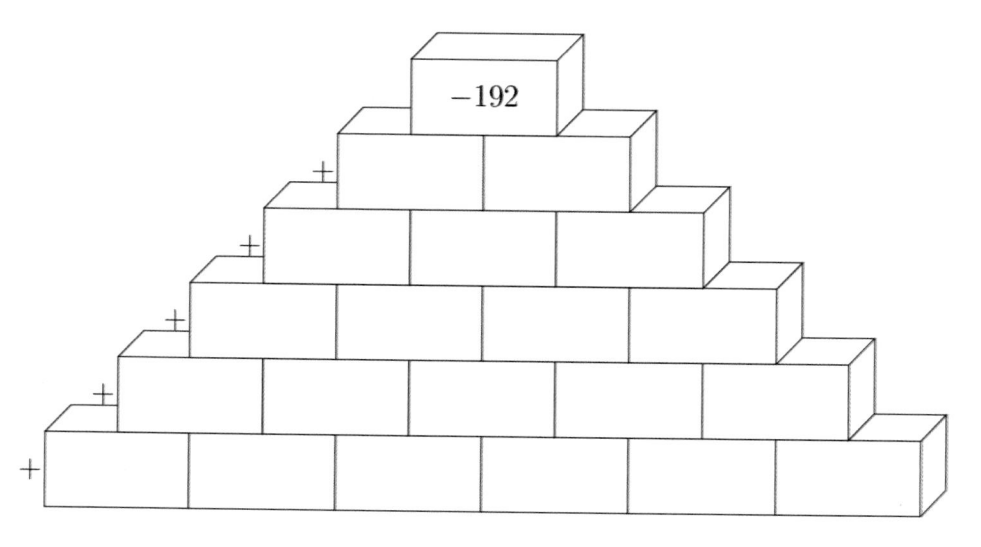

8. Find your way through the maze of numbers. The "signpost" to the right of the exercise shows which direction you may go. For example: you can only go down if the number in the cell below is 9 larger than the number of the current cell. There is only one direction consistent with the signpost for each step. Mark your path on the figure.

Start

−2	−12	−2	−10	−2
−13	−1	+9	−1	+7
−28	−16	−4	+8	−2
−19	−27	+11	−19	−27
+8	−18	+20	−28	−18

Finish

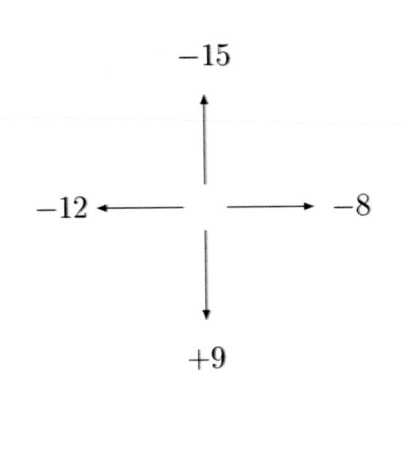

9. In a game, Sebastian has 56 points less than Eve. Eve has 38 points, Sebastian also gets an additional 15 points. How many points does Sebastian have and how far behind Eve is he then?

10. Fill in the empty spaces.

Here one uses addition:

+	+4,3	−1,9	+5,6
−7,4			
+2,8			
−3,1			

+			+10
−6		+1	
			−7
	−4	0	

In the following two tables subtract the number in the gray column from those in the grey row.

−	+4,3	−1,9	+5,6
−7,4			
+2,8			
−3,1			

−			+10
−6		+1	
			−7
	−4	0	

11. Replace the unknown x with a number to make the calculation correct.

    a) (-4) minus $x = (+13)$ $x =$

    b) x minus $(-4) = (-3)$ $x =$

    c) $(+5)$ minus $x = (+10)$ $x =$

    d) x minus $(+22) = (-33)$ $x =$

    e) $2 \cdot x$ minus $(-4) = (+10)$ $x =$

12. Calculate:

    a) Which number do you have to subtract five times from 5 to get -30?

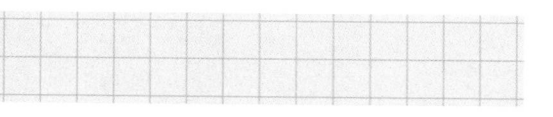

    b) Which number do you have to subtract six times from 6 to get -60?

    c) Which number do you have to subtract seven times from -7 to get -63?

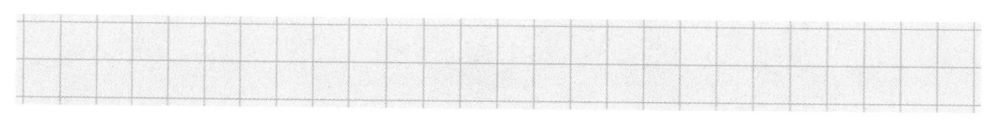

    d) Which number do you have to add eight times to -8 to get 64?

e) Which number do you have to add nine times to -90 to get -36?

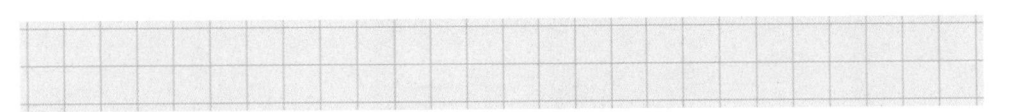

3.3 Simplifying notation

1. Simplify the written form of the calculation and find the answer.

Example: $\boxed{(-10) + (-15) = -10 - 15 = -25}$

a) $(+23) - (+45) =$

b) $(-3) + (+4) =$

c) $(-10) + (-6) =$

d) $(+4) - (-3) =$

e) $(+2) - (+5) =$

f) $(-3) - (-3) =$

g) $(-23) + (+13) =$

h) $(-12) - (+25) =$

i) $(+15) + (+4) =$

j) $0 - (-3) =$

k) $(-3) - 0 =$

l) $(+4) - (+2{,}3) =$

m) $(+16) - (-32) =$

n) $(-42) - (-53) =$

o) $(-52) - (+41) =$

2. Simplify the notation and find the answer.

   a) $(+25) - (-2) + (+12) - (+35) + (-21) - (-5) =$

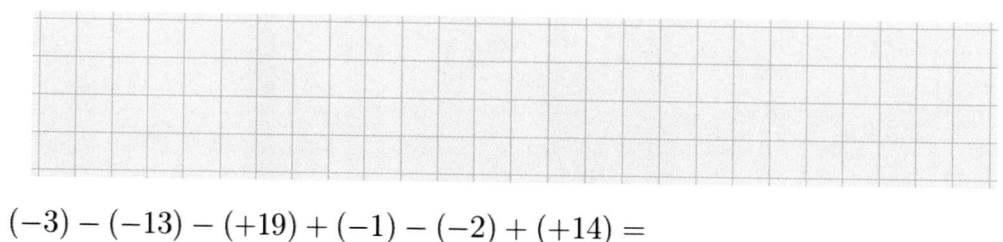

   b) $(-3) - (-13) - (+19) + (-1) - (-2) + (+14) =$

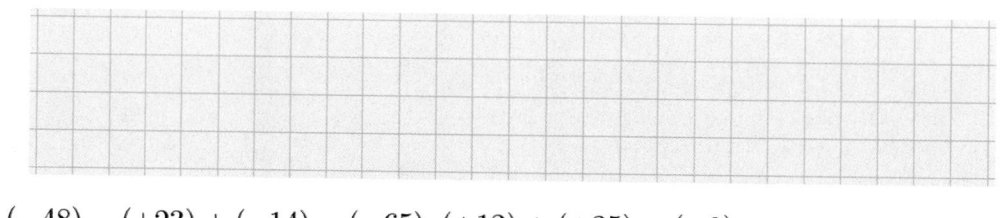

   c) $(-48) - (+23) + (-14) - (-65) - (+12) + (+25) - (+9) =$

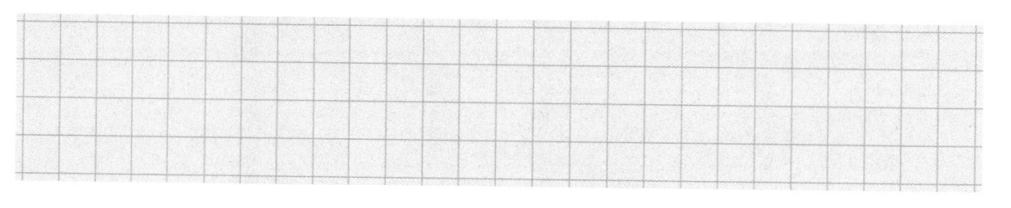

3. Calculate efficiently; state how you got your answer.

   a) $-3 + 8 - 47 + 12 =$

   b) $-12 + 87 + 24 - 12 + 33 =$

   c) $-4 - 19 + 6 + 4 + 9 =$

   d) $-24 + 49 + 63 - 26 =$

   e) $22 + (-18) + (-7) + 8 + (-25) =$

   f) Invent three of your own exercises.

4. Calculate:

$$-2 + 3 - 4 + 5 - 6 + 7 - 8 =$$

Reverse the calculation signs and find the answer. What do you notice?

5. Simplify and calculate:

$$(-1) - (-2) + (-3) - (-4) + (-5) - (-6) + (-7) - (-8) + (-9) - (-10) =$$

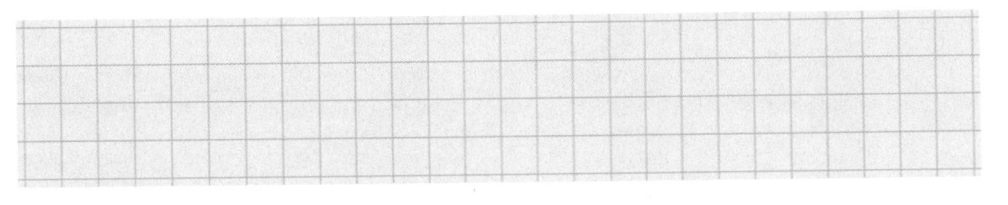

What would be the results if you continue series further?

   a) up to $\ldots + (-19) - (-20)$?

   b) up to $\ldots + (-35)$?

   c) up to $\ldots - (-100)$?

6. Simplify and calculate:

$$1 - (-3) - (-5) - (-7) - (-9) =$$

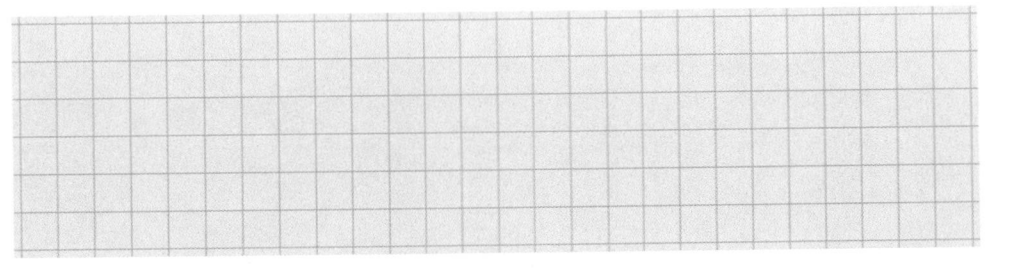

What would be the result, if you carry on up to:

a) $\ldots - (-21)$?

b) up to $\ldots - (-49)$?

c) up to $\ldots - (-101)$?

7. In the following number pyramids some arithmetic operations are given but some are also missing. Fill in the blank spaces (both missing numbers and missing operation signs). Sometimes there are more than one possibility to fill in correctly.

example: possible solutions:

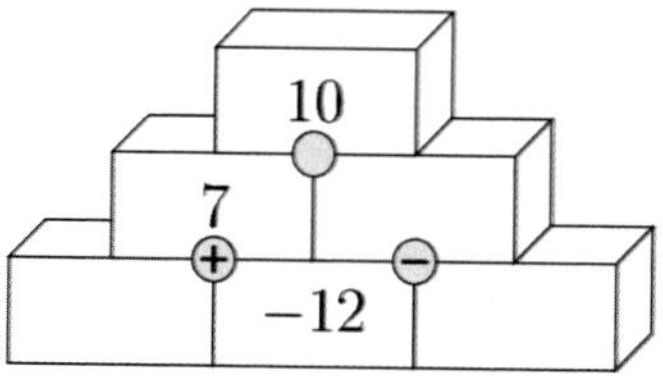

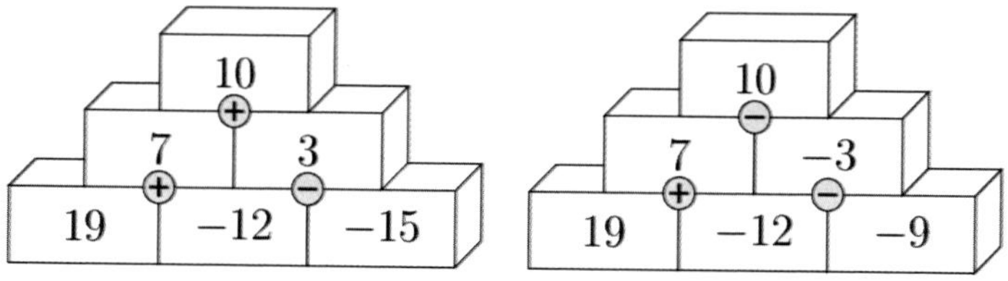

a)

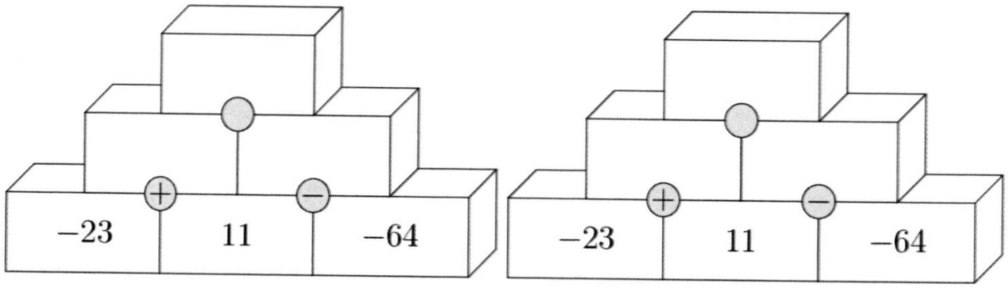

b)

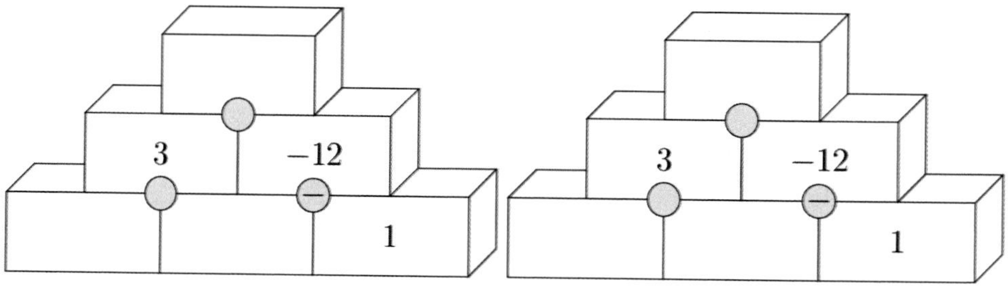

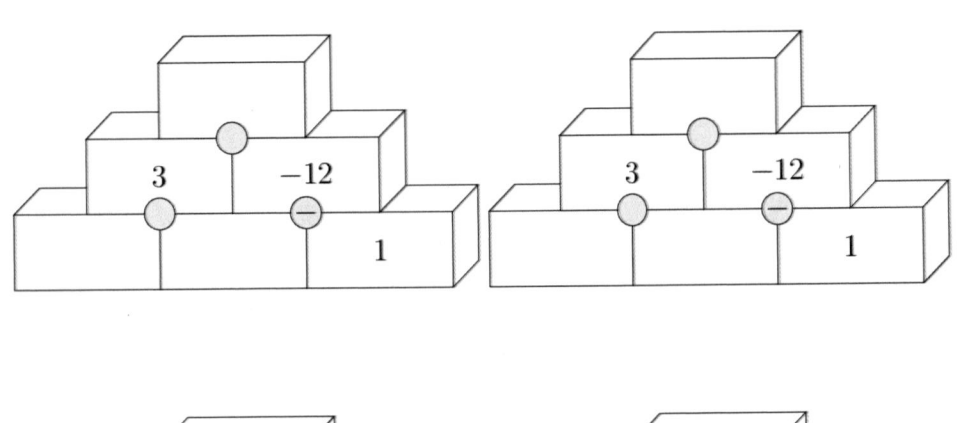

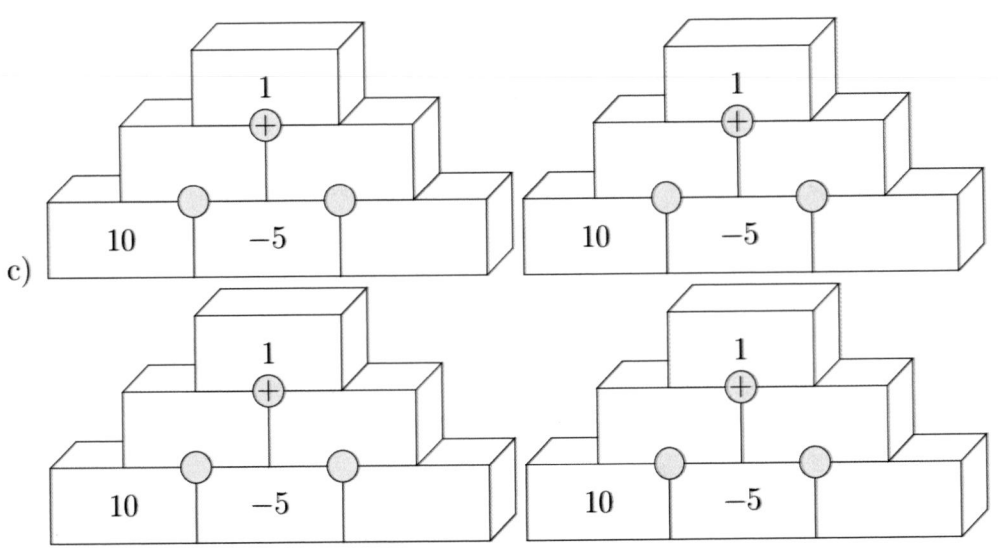

c)

8. You are given seven numbers, three positive: $+2$, $+7$, $+11$ and four negative: -4, -13, -20 -1.

a) Form the sum.

b) Decrease each of the negative numbers by 9, increase each of the positive numbers by 9. Now calculate the sum of the numbers. What do you notice?

   c) Decrease all the positive numbers by 9, increase all the negative numbers by 9. What happens now?

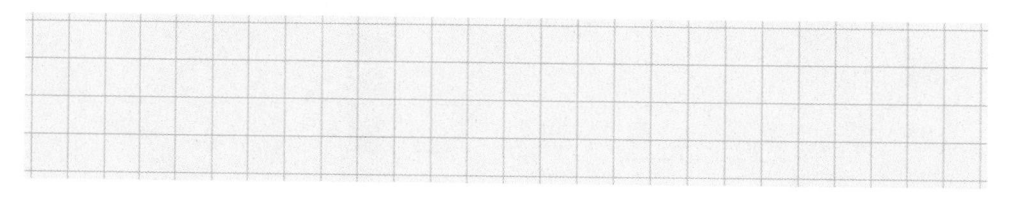

9. $12675 - 55489 =$

   a) Make a rough calculation, then make an exact calculation and round up to the nearest thousand.

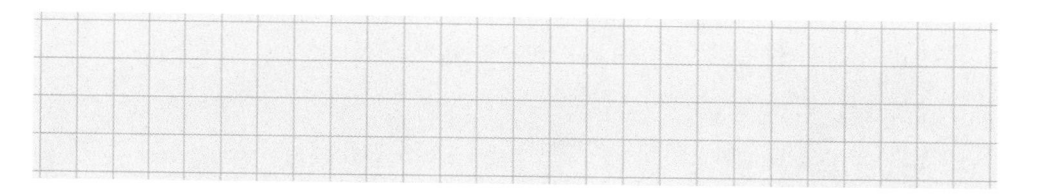

   b) Omit 2 digits in the first term so that the answer is as small as possible. What answer do you obtain?

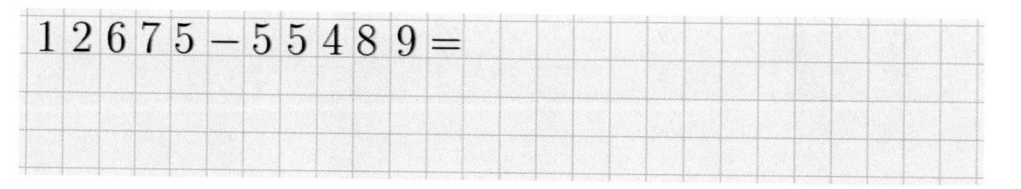

   c) Omit 2 digits in the second term so that the answer is as small as possible. What is your answer?

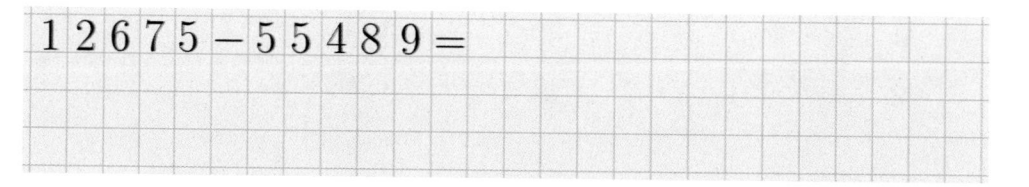

   d) Omit a total of 4 digits in the two terms so that the resulting answer is as large as possible. Share the answer you obtain?

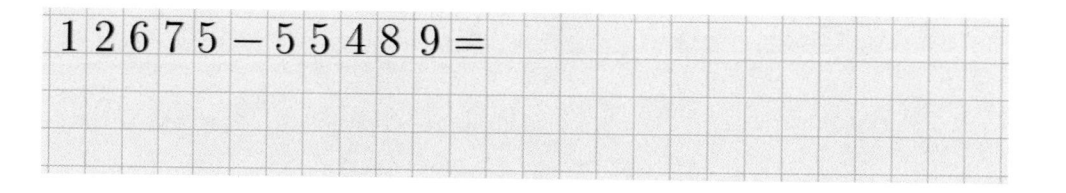

3.4 Parentheses and precedence rules

1. Calculate the innermost parentheses first

   Example:

   $$(-12) + \big((+25) + (-24)\big) = -12 + \underbrace{(25 - 24)}_{1} = -12 + 1 = -11$$
   $$\underbrace{}_{25-24}$$

   a) $(-4) - \big((-12) + (+13)\big) = $ _____

   b) $(-5) + \big((-22) - (-22)\big) = $ _____

   c) $(+7) + \big((-23) + (+11)\big) = $ _____

   d) $\big((-12) - (+23)\big) + (+30) = $ _____

   e) $(-17) - \big((-17) - (-5)\big) = $ _____

2. Always calculate the parentheses first (when they are present).

   a) $-13 + (12 - 5) = $ _____

   b) $(-13 + 12) - 5 = $ _____

   c) $-13 + 12 - 5 = $ _____

   d) $-13 - (12 - 5) = $ _____

   e) $(-13 - 12) - 5 = $ _____

   f) $-13 - 12 - 5 = $ _____

   g) Which parentheses from the six exercises could you leave out without influencing the answer?

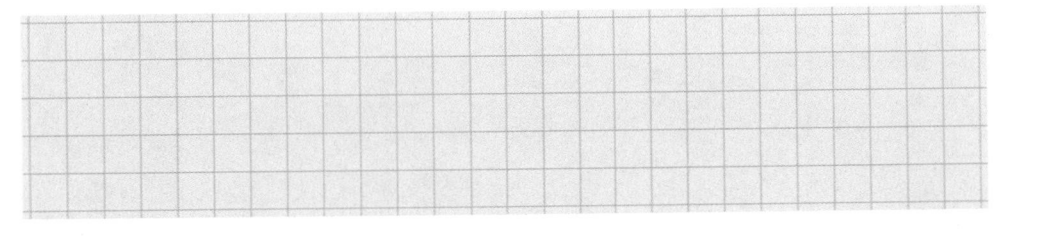

3. Calculate the numbers in the parentheses first:

    a) $14 - (32 + 3) =$ _____

    b) $-14 - (32 - 3) =$ _____

    c) $14 - (-32 + 3) =$ _____

    d) $-14 + (32 - 3) =$ _____

    e) $14 - (-32 - 3) =$ _____

    f) $14 + (-32 - 3) =$ _____

    g) $14 - (32 - 3 - 2) =$ _____

For every exercise try to obtain the same answers using the same numbers without using parenthesis but rather by changing the numerical signs.

Example: $\mathbf{14 - (32 + 3)} = 14 - 35 = -21 = \mathbf{14 - 32 - 3}$

4. Calculate the parentheses in two different ways as in the example and continue the series for three further rows.

Example:

$$2 \cdot \underbrace{(10 + 3)}_{13} = 2 \cdot 13 = 26$$
$$2 \cdot (10 + 3) = 2 \cdot 10 + 2 \cdot 3 = 20 + 6 = 26$$

a) $7 \cdot (10 + 5) =$ $7 \cdot (10 + 5) =$

b) $7 \cdot (20 + 5) =$ $7 \cdot (20 + 5) =$

c) $7 \cdot (30 + 5) =$ $7 \cdot (30 + 5) =$

d) $7 \cdot (40 + 5) =$ $7 \cdot (40 + 5) =$

e)

f)

g)

h) How does the calculation in the 10th row turn out?

i) What is the answer in the 20th row?

5. Now calculate the parentheses as in exercise 4 in two ways and continue the sequence for three further rows.

a) $3 \cdot (10 + 2) =$ $3 \cdot (10 + 2) =$

b) $4 \cdot (20 + 2) =$ $4 \cdot (20 + 2) =$

c) $5 \cdot (30 + 2) =$ $5 \cdot (30 + 2) =$

d) $6 \cdot (40 + 2) =$ $6 \cdot (40 + 2) =$

e)

f)

g)

4 Multipliying positive and negative numbers

4.1 Introductory exercises

1. Calculate the parentheses in two different ways as given in the example. Continue the sequence for three further rows.

Example:

$$(\underbrace{10 - 3}_{7}) \cdot 5 = 7 \cdot 5 = 35$$
$$(10 - 3) \cdot 5 = 50 - 15 = 35$$

a) $(10 - 3) \cdot 4 =$ $(10 - 3) \cdot 4 =$

b) $(10 - 3) \cdot 3 =$ $(10 - 3) \cdot 3 =$

c) $(10 - 3) \cdot 2 =$ $(10 - 3) \cdot 2 =$

d) $(10 - 3) \cdot 1 =$ $(10 - 3) \cdot 1 =$

e) _____

f) _____

g) _____

2. Complete the multiplication table. Write negative numbers in red.

·	5	4	3	2	1	0	−1	−2	−3	−4	−5
5	25										
4											
3			6								
2											
1											
0											
−1											
−2											
−3											
−4											
−5											

3. Calculate:

a) $2 \cdot (-14) =$ _____

b) $(-2) \cdot (-14) =$ _____

c) $(-14) \cdot 4 =$ _____

d) $(-14) \cdot (-4) =$ _____

e) $(-7) \cdot (-8) =$ _____

f) $(-24) \cdot 38 =$ _____

g) $24 \cdot (-19) =$ _____

h) $(-8) \cdot (-19) =$ _____

i) $(-24) \cdot (-190) =$ _____

j) $240 \cdot (-19) =$ _____

4. Write the given number as product of two factors in as many different ways as you can:

a) -36

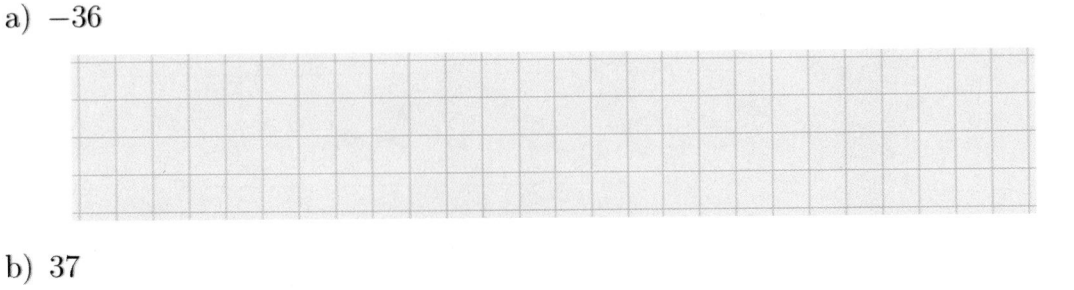

b) 37

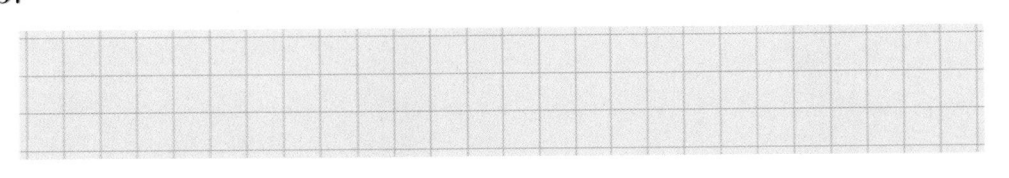

How many are there of each?

5. Calculate and multiply as efficiently as possible. Sometimes it may be necessary to calculate on a separate sheet of paper.

a) $-3 \cdot (-4) \cdot (-5) =$

b) $(-5) \cdot 14 \cdot (-2) =$

c) $6 \cdot (-2) \cdot (-3) =$

d) $25 \cdot (-2) \cdot 16 \cdot (-2) =$

e) $12 \cdot 3 \cdot (-5) =$

f) $-17 \cdot (-21) \cdot (-12) =$

g) $(-2) \cdot (-1) \cdot (-1) \cdot (-1) =$

Without calculating, how can you see whether the answer is positive or negative?

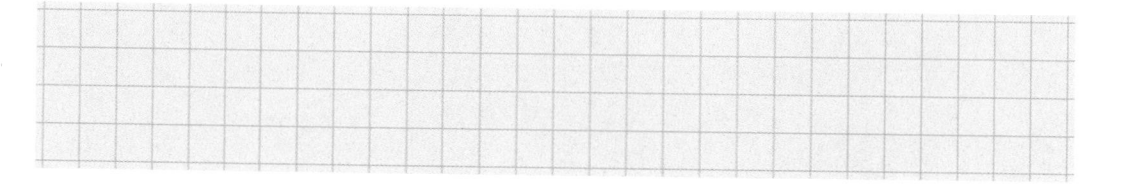

6. How many products are there of two of the numbers $-12, \ -11, \ -10, \ldots, \ 11, \ 12$

a) are larger than 100,

b) are smaller than or equal to $+4$ and simultaneously larger or equal to -4?

7. Which numbers are missing in the parentheses?

  a) $(-6) \cdot (\quad) = -36$

  b) $(-5) \cdot (\quad) = 30$

  c) $(\quad) \cdot (-13) = -52$

  d) $(-14) \cdot (\quad) = 196$

  e) $3 \cdot (\quad) = -3$

  f) $17 \cdot (\quad) = -289$

8. Susi believes that the product of two numbers is always larger than either of the two factors.

  a) Explain using the product of 2 multiplied by 5, what is meant by the above sentence.

  b) Find examples with whole numbers (no fractions!) in which this "rule" is broken.

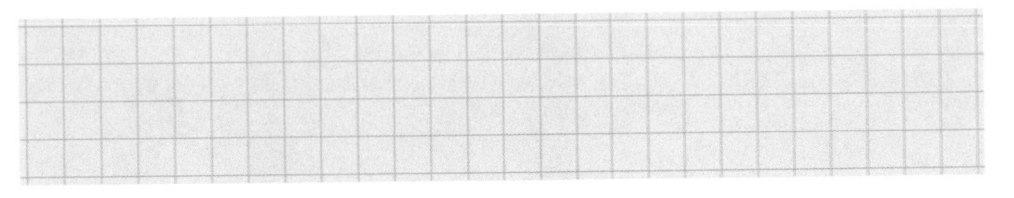

  c) How should one formulate the rule correctly?

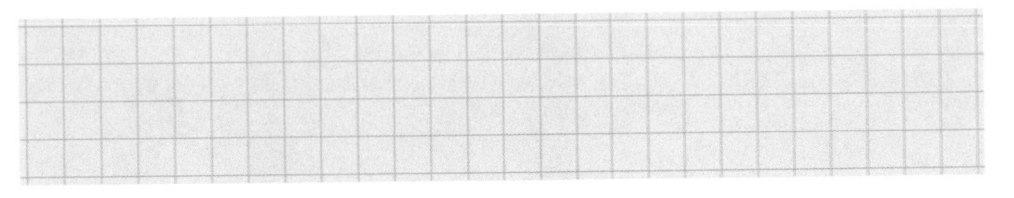

9. The following is a multiplication pyramid. Each brick contains the product of the numbers in the two bricks it is resting on.

a) Fill in the blank bricks.

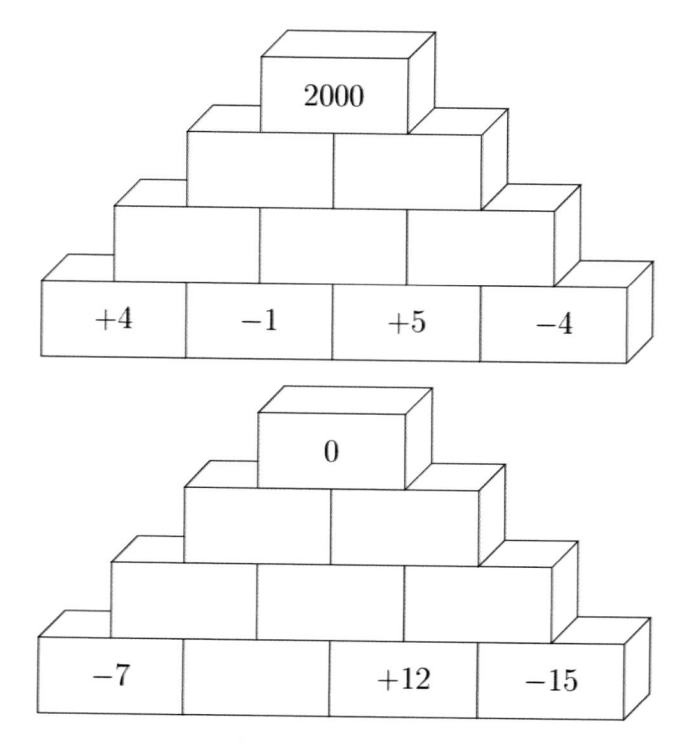

b) How does the number at the top change when the numbers at the bottom are

- doubled?

- multiplied by (-1)?

c) Imagine a pyramid with six layers. What sign would the top number have if the numbers at the bottom

- were alternating positive and negative numbers

- are all negative?

— consist of 2 positive and 4 negative numbers. Does it matter which bricks in this row are negative?

10. You learned what a magic square is in exercise 1 (on page 24). Below you see a numbered square which is clearly not an ordinary magic square.

64	−128	−4
−2	32	−512
−256	−8	16

a) Check whether the square is also a magic "multiplication square".

b) Can you change the magic multiplication square to produce another magic multiplication square?

c) Write all the numbers of the square in ascending order without considering the signs. What do you notice? Can you also make a magic square with completely different numbers?

11. Decide whether the following statements are true.

    a) $3100 \cdot 74 > 3101 \cdot 73$

    b) $24 \cdot 12589 > 27 \cdot 12588$

    c) $(-3100) \cdot 74 > 3101 \cdot (-73)$

    d) $(-24) \cdot 12589 > (-27) \cdot 12588$

4.2 Multiplication by powers of 10

1. Calculate the following exercises and formulate a rule for multiplication by 10, 100, 1000, ... and by 0.1; 0.01; 0.001 ... .

$$(-48) \cdot 0.0001 = \underline{\hspace{6cm}}$$

$$(-48) \cdot 0.001 = \underline{\hspace{6cm}}$$

$$(-48) \cdot 0.01 = \underline{\hspace{6cm}}$$

$$(-48) \cdot 0.1 = \underline{\hspace{6cm}}$$

$$(-48) \cdot 1 = \underline{\hspace{6cm}}$$

$$(-48) \cdot 10 = \underline{\hspace{6cm}}$$

$$(-48) \cdot 100 = \underline{\hspace{6cm}}$$

$$(-48) \cdot 1000 = \underline{\hspace{6cm}}$$

$$(-48) \cdot 10000 = \underline{\hspace{6cm}}$$

2. Calculate the following and think about how you can use the rule from exercise 1 on the facing page.

$$(-4) \cdot 0.0012 = $$

_____

$$(-4) \cdot 0.012 = $$

_____

$$(-4) \cdot 0.12 = $$

_____

$$(-4) \cdot 1.2 = $$

_____

$$(-4) \cdot 12 = $$

_____

$$(-4) \cdot 120 = $$

_____

$$(-4) \cdot 1200 = $$

_____

$$(-4) \cdot 12000 = $$

_____

$$(-4) \cdot 120000 = $$

_____

3. Calculate

a) $1.75 \cdot (-1) =$ e) $17500 \cdot 0.001 =$

b) $(-17.5) \cdot (-100) =$ f) $1000 \cdot (-1.75) =$

c) $175 \cdot (-0.1) =$ g) $0.175 \cdot 10 =$

d) $(-1.75) \cdot 10 =$ h) $-0.1 \cdot 0.175 =$

4. Calculate

a) $(-3) \cdot 4 =$ f) $0.3 \cdot (-40) =$

b) $0.3 \cdot (-4) =$ g) $(-0.03) \cdot (-400) =$

c) $(-0{,}3) \cdot 0.4 =$ h) $0.004 \cdot 3000 =$

d) $40 \cdot 3 =$ i) $0.3 \cdot (-0.4) =$

e) $40 \cdot (-30) =$ j) $-0.03 \cdot (-0.04) =$

5. Calculate

a) $(-2) \cdot 0.3 = $ _____

b) $(-12) \cdot (-8) = $ _____

c) $0.12 \cdot 8 = $ _____

d) $1.1 \cdot (-1.1) = $ _____

e) $(-0.7) \cdot (-7) = $ _____

f) $0.3 \cdot 0.3 = $ _____

g) $(-4) \cdot $ _____ $= -1,2$

h) $10000 \cdot $ _____ $= -1000$

i) $(-1.2) \cdot 0.5 = $ _____

j) $2 \cdot $ _____ $\cdot (-0.5) = 5$

k) $(-1) \cdot (-1) \cdot (-1) \cdot $ _____ $= 10$

l) $-50 \cdot 0.2 = $ _____

m) $40 \cdot (-0.25) = $ _____

n) $4 \cdot (-0.75) = $ _____

4.3 Multiplication by fractions

1. Calculate and convert the result into decimal numbers. Reflect on how the exercises are connected to exercises 1 and 2 on page 53. Convert all answers to decimal numbers, then rewrite each exercise using decimal numbers only.

a) $\left(-\dfrac{1}{10}\right) \cdot 3 =$ _____

b) $\left(-\dfrac{1}{100}\right) \cdot (-30) =$ _____

c) $\dfrac{1}{100} \cdot (-3) =$ _____

d) $\left(-\dfrac{3}{10}\right) \cdot \dfrac{1}{10} =$ _____

e) $\dfrac{1}{1000} \cdot (-300) =$ _____

f) $\left(-\dfrac{3}{10}\right) \cdot \left(-\dfrac{3}{10}\right) =$ _____

g) $-3 \cdot \dfrac{3}{100} =$ _____

h) $3000 \cdot \left(-\dfrac{3}{10}\right) =$ _____

2. Calculate. Don't forget to reduce to simplest form!

a) $\dfrac{2}{3} \cdot \dfrac{3}{5} =$ _____

b) $\dfrac{3}{5} \cdot \left(-\dfrac{4}{9}\right) =$ _____

c) $\left(-\dfrac{2}{3}\right) \cdot \left(-\dfrac{2}{3}\right) =$ _____

d) $\left(-\dfrac{8}{35}\right) \cdot \dfrac{49}{24} =$ _____

e) $\dfrac{3}{8} \cdot \left(-\dfrac{24}{5}\right) \cdot \dfrac{25}{3} =$ _____

f) $\dfrac{4}{5} \cdot$ _____ $= -\dfrac{8}{15}$

4.4 Mixed exercises

1. Calculate the parentheses in two different ways as given in the example. Continue the sequence for three more rows.

   Example:

   $$2 \cdot (\underbrace{10 - 3}_{7}) = 2 \cdot 7 = 14$$
   $$2 \cdot (10 - 3) = 2 \cdot 10 - 2 \cdot 3 = 20 - 6 = 14$$

   a) $7 \cdot (10 - 5) =$ $7 \cdot (10 - 5) =$

   b) $7 \cdot (20 - 5) =$ $7 \cdot (20 - 5) =$

   c) $7 \cdot (30 - 5) =$ $7 \cdot (30 - 5) =$

   d) $7 \cdot (40 - 5) =$ $7 \cdot (40 - 5) =$

   e)

   f)

   g)

   h) Show the calculation in the 10th row.

   i) What is the answer in the 20th row?

2. Calculate the parentheses in two different ways as given in the first exercise. Continue the sequence for three more rows.

a) $3 \cdot (10 - 2) =$ $3 \cdot (10 - 2) =$

b) $4 \cdot (20 - 2) =$ $4 \cdot (20 - 2) =$

c) $5 \cdot (30 - 2) =$ $5 \cdot (30 - 2) =$

d) $6 \cdot (40 - 2) =$ $6 \cdot (40 - 2) =$

e)

f)

g)

3. Calculate as in the example.

Example:

$$14 - \underbrace{(-2) \cdot 3}_{=(-6)} = 14 - (-6) = 14 + 6 = 20$$

a) $17 - 8 \cdot (-3) =$

b) $18 \cdot (-5) + 32 =$

c) $3 \cdot 7 - (-4) \cdot (-9) =$

d) $25 + (-26) \cdot (-27) =$

e) $25 - 26 \cdot 27 =$

f) $25 - (-26) \cdot 27 =$

g) $(25 - 26) \cdot 27 =$

h) $(25 - 26) \cdot (-27) =$

4. Calculate and continue for three further rows. Try to justify the pattern shown in the answers.

a) $2 \cdot (-3) \ + \ 3 \cdot (-2) \ =$ _____

$3 \cdot (-3) \ + \ 2 \cdot (-2) \ =$ _____

$4 \cdot (-3) \ + \ 1 \cdot (-2) \ =$ _____

_____

_____

_____

b) $5 \cdot 1 \ - \ 4 \cdot 2 \ + \ 3 \cdot 3 \ =$ _____

$5 \cdot 2 \ - \ 4 \cdot 3 \ + \ 3 \cdot 2 \ =$ _____

$5 \cdot 3 \ - \ 4 \cdot 4 \ + \ 3 \cdot 1 \ =$ _____

_____

_____

_____

5. Continue the pattern of calculation for a further three rows. What do you notice about the answers?

$(-3) \cdot (+4) + (3) \cdot (-4) - (+4) \cdot (+4) \ =$ _____

$(-2) \cdot (+5) + (4) \cdot (-3) - (+3) \cdot (+3) \ =$ _____

$(-1) \cdot (+6) + (5) \cdot (-2) - (+2) \cdot (+2) \ =$ _____

_____

_____

_____

6. Calculate.

a) $(+1) \cdot (+2) \cdot (+3) - (+1) \cdot (-2) \cdot (+3) + (-1) \cdot (+2) \cdot (+3) + (+1) \cdot (+2) \cdot (-3) =$

b) $6 - 2 \cdot (-3) - 1 \cdot (-2) \cdot 3 - (-1) \cdot 2 \cdot (-3) + 1 \cdot 2 \cdot (-3) =$

7. Simplify, calculate and complete the last three rows.

a) $(-3) \cdot \big(2 + (-3)\big) =$

   $(-2) \cdot \big(3 + (-2)\big) =$

   $(-1) \cdot \big(4 + (-1)\big) =$

b) $\big(4 + (-2)\big) \cdot \big(3 - (+2)\big) =$

   $\big(3 + (-2)\big) \cdot \big(2 - (+2)\big) =$

   $\big(2 + (-2)\big) \cdot \big(1 - (+2)\big) =$

c)
$$\big(1 + (-2)\big) \cdot \big((0) - (+2)\big) = \underline{\hspace{5cm}}$$

$$\big(0 + (-2)\big) \cdot \big((-1) - (+2)\big) = \underline{\hspace{5cm}}$$

$$\big((-1) + (-2)\big) \cdot \big((-2) - (+2)\big) = \underline{\hspace{5cm}}$$

_____

_____

_____

8. Find two numbers

a) whose sum is -11 and whose product is $+30$.

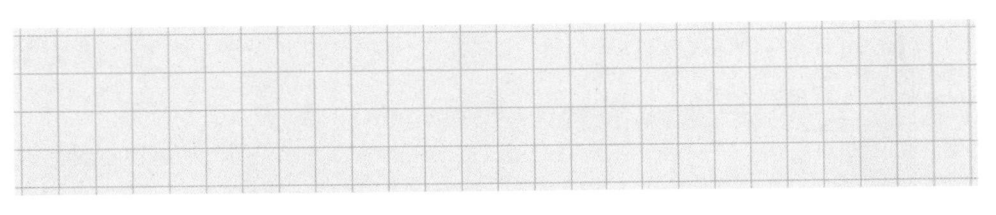

b) whose sum is -1 and whose product is -30,

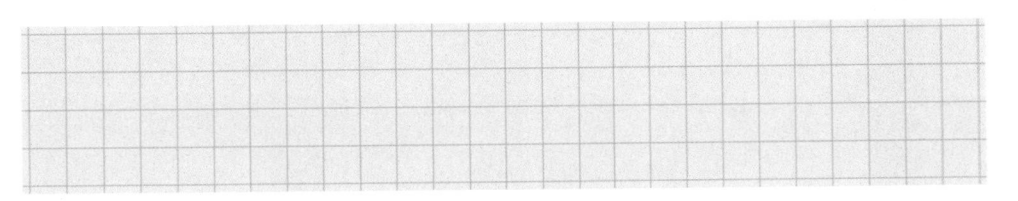

c) whose sum is $+1$ and whose product is -30,

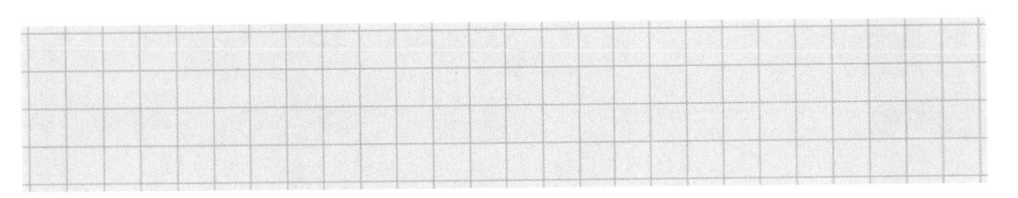

d) whose sum is -25 and whose product is $+24$.

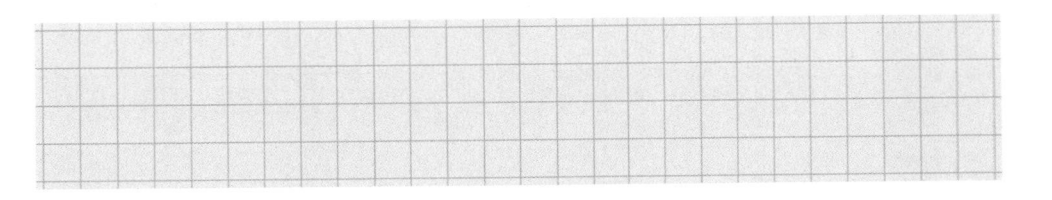

9. Read the exercises very carefully and and for each one attempt to find a fitting expression to calculate.

   a) Calculate the difference of the product of 17 and 4 and the number -38.

   b) Calculate the product of the difference of 17 and 4 and the number -38.

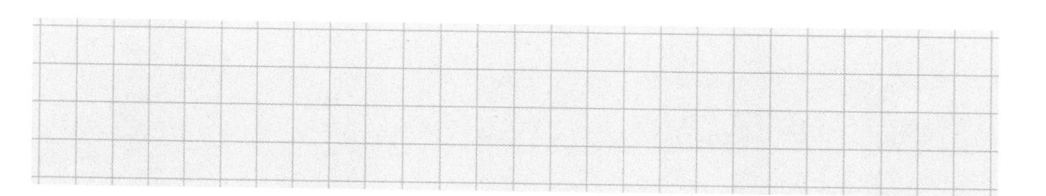

   c) Calculate the sum of the product and the difference of the numbers -4 and -38.

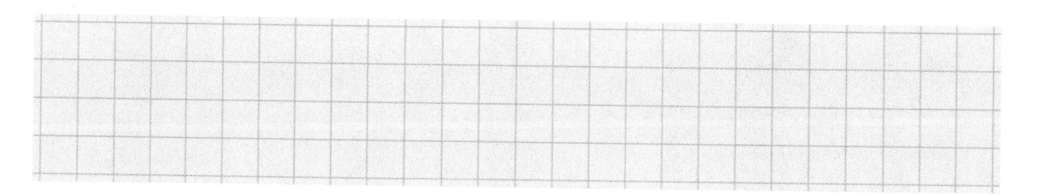

10. Find signs and operations, and possibly parentheses, which when placed between the numbers yield the given answer. It often happens that there are several possibilities to do this.

    Example:

5	3		1 = -10	becomes $5 \cdot (-3 + 1) = -10$

    a) 4   1   5 = 25   d) 4   1   5 = 16

    b) 4   1   5 = -16   e) 4   1   5 = -15

    c) 4   1   5 = -2   f) 4   1   5 = 1

11. Complete as in the previous exercise.

    a) 5   3   2   1 = 4

    b) 4   7   5   1 = -12

    c) 3   2   4   5 = -21

4.5 Some brain teasers

Here a three digit number is multiplied by a two digit number. Each letter represents a number between 0 and 9. Different letters stand for different digits.

a)

```
    A  4  C  ·  A  D
    _____
          4  E  A
       D  D  D
    _____
       4  E  A  D
```

b)

```
    F  2  G  ·  F  F
    _____
       F  2  G
       F  2  G
    _____
       F  G  H  G
```

c)

```
    I  N  U  ·  N  U
    _____
       N  U  S
       L  N  U
    _____
    O  I  N  U
```

5 Dividing positive and negative numbers

5.1 Introductory exercises

1. Calculate:

a) $(-21) \div 7 =$ ⬚

b) $56 \div (-8) =$ ⬚

c) $(-72) \div (-18) =$ ⬚

d) $(-256) \div 16 =$ ⬚

e) $1024 \div (-4) =$ ⬚

f) $(-125) \div (-12{,}5) =$ ⬚

2. Fill in the gaps:

a) _____ $\div (-4) = 6$

b) $36 \div$ _____ $= -4$

c) _____ $= 45 \div (-9)$

d) $(-243) \div$ _____ $= 81$

e) _____ $\div (-9) = -7$

f) $(-360) \div$ _____ $= 18$

g) _____ $= (-32) \div 8$

h) $4.5 \div$ _____ $= -1.5$

i) _____ $\div (-2.5) = 16$

j) _____ $\div (-4) = 0.75$

3. Calculate and write the answer in decimal form:

a) $\dfrac{15}{-5} =$ _____

b) $\dfrac{-120}{12} =$ _____

c) $\dfrac{-42}{-6} =$ _____

d) $\dfrac{121}{-11} =$ _____

e) $\dfrac{-6561}{-9} =$ _____

f) $\dfrac{-32}{10} =$ _____

g) $\dfrac{48}{-10} =$ _____

h) $\dfrac{48}{-1{,}2} =$ _____

i) $\dfrac{48}{1{,}2} =$ _____

j) $\dfrac{-4{,}8}{-1{,}2} =$ _____

k) $\dfrac{-4{,}8}{12} =$ _____

l) $\dfrac{-21}{-14} =$ _____

m) $\dfrac{96}{-8} =$ _____

n) $\dfrac{-420}{12} =$ _____

o) $\dfrac{-384}{-15} =$ _____

p) $\dfrac{124{,}8}{-5} =$ _____

4. Give all the quotients that can be written using the numbers -7, 168, 41, 0, -12, 35 that produce whole-number answers (without remainders) and calculate their value..

5. You can make many quotients from the numbers -48, -32, -24, -3, 2, 6, 15 and 30 . The quotients don't have to yield a whole number (integer). Find the quotient

    a) with the smallest value:

    b) with the largest value:

    c) with the nearest value to zero:

5.2 Division by powers of 10

1. $(-48) \div 0.0001 =$

   _____

   $(-48) \div 0.001 =$

   _____

   $(-48) \div 0.01 =$

   _____

   $(-48) \div 0.1 =$

   _____

   $(-48) \div 1 =$

   _____

   $(-48) \div 10 =$

   _____

   $(-48) \div 100 =$

   _____

   $(-48) \div 1000 =$

   _____

   $(-48) \div 10000 =$

   _____

   a) Calculate. Formulate rules for division by 10, 100, 1000 ... and by 0.1; 0.01; 0.001 ... .

b) Compare the results with exercise 1 on page 52. What do you notice? Can you explain it?

2. Calculate. Compare with the calculations in exercise 2 on page 53.

$$(-48) \div 0.0012 = $$

$$(-48) \div 0.012 = $$

$$(-48) \div 0.12 = $$

$$(-48) \div 1.2 = $$

$$(-48) \div 12 = $$

$$(-48) \div 120 = $$

$$(-48) \div 1200 = $$

$$(-48) \div 12000 = $$

$$(-48) \div 120000 = $$

6 Mixed exercises

1. Calculate.

a) $-17 + 63 \div (-7) =$ ⎯⎯⎯⎯⎯⎯⎯⎯⎯⎯⎯⎯

b) $400 - (-39) \div (-3) =$ ⎯⎯⎯⎯⎯⎯⎯⎯⎯⎯⎯⎯

c) $4 + 5 \cdot (-6) =$ ⎯⎯⎯⎯⎯⎯⎯⎯⎯⎯⎯⎯

d) $17 + \left(-38 + \left(8 - 24 \div (-8)\right)\right) =$ ⎯⎯⎯⎯⎯⎯⎯⎯⎯⎯

e) $(-800) \div 25 + 200 =$ ⎯⎯⎯⎯⎯⎯⎯⎯⎯⎯⎯⎯

f) $35 \div (15 - 22) =$ ⎯⎯⎯⎯⎯⎯⎯⎯⎯⎯⎯⎯

g) Describe the arithmetic expressions of the first three examples in your own words:

2. Convert to an arithmetic expression and calculate..

a) Add the quotient of 63 and -7 to -17.

b) Divide the sum of 35 and (-11) by -8.

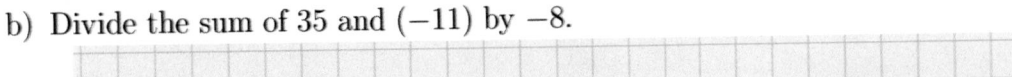

c) Find the difference of 7 and -43 and multiply by 10.

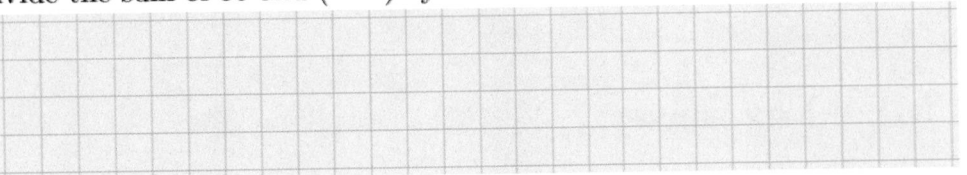

d) Find the difference of the quotient of -12 and -2 and the product of the same numbers.

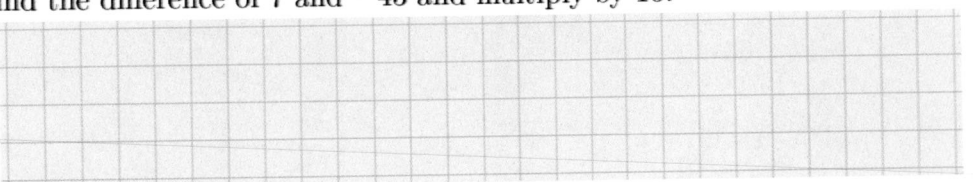

3. Fill in the blanks.

a) $-5+$ _____ $= -13$

b) $2 \cdot (-17) =$ _____

c) $72 = (8) \cdot$ _____

d) _____ $\div 14 = -5$

e) $12 = 25 +$ _____

f) $-4 =$ _____ -8

g) $(-0.1) \cdot 12{,}5 =$ _____

h) $-21-$ _____ $= 4$

i) _____ $\cdot (-21) = 126$

4. Fill in the correct sign in the boxes.

    a) $8 \,\square\, (-6) \,\square\, (-2) = -4$

    b) $(-15) \,\square\, (-3) \,\square\, 4 = 1$

    c) $(-2) \,\square\, 1 \,\square\, (-3) = 1$

    d) Now create your own exercise and give it to your neighbour.

    e) Find exercises of this kind with more than one solution.

5. Put in ">" , "<" or "=" so that the statement is correct. Justify your decision without answering the terms:

    a) $(-408) \div 17 \, \square \, 408 \div (-17)$

    b) $1112 \div (-4) \, \square \, (-1112) \div 5$

    c) $9545 \div (-415) \, \square \, (-9546) \div 415$

6. Simplify the arithmetic expressions and then find out the value.

    a) $\left((-45) \div (-3) + 10 \div (-2)\right) \div (-5) =$

    b) $\left(18 \div (-2) - 90 \div (-3)\right) \div (-7) =$

    c) $(18 - 8 \div 2 \cdot 2) \div \left(-4 + 4 \cdot (-2)\right) =$

    d) $\left(-45 \div 9 + 5 \cdot 32 \div (-8)\right) \div (-5) \cdot (-5) =$

Made in the USA
Las Vegas, NV
20 September 2023

77836344R00043